Hello and Welcome to the

Arduino Robot Building Workshop!

Module 1

This is the first of a 10 part series in building and programming Arduino robots. By the time you complete all 10 modules you should have yourself a little Arduino robot running around the floor, beeping, blinking, following lines, entering sumobot contests, and more. This is a beginner's course. It starts with the easiest concepts and goes on to building complete robots. Please don't just read this course please make sure you DO THE LABS. You learn by doing, so do a lot and learn a lot!

© 2012 Robots-and-Androids.com

VERY IMPORTANT NOTE

Please note that this is my first effort at writing an eBook. As it is my first effort you are bound to find errors and issues with this course, so **PLEASE LET ME KNOW WHEN YOU FIND MISTAKES** and I will correct them.

We have a blog on our site in which to report errors, and we will put corrections immediately on our web site (http://www.robots-and-androids.com) and periodically into new releases after you send them to us.

HAPPY ROBOT BUILDING!

This Course Assumes:

> - You are over 12 (under 18 MUST be supervised by a parent.)
> - You own a working PC (Windows XP or better.)
> - You know your way around a PC (cut, copy, paste, open files, save files, etc.)
> - You have all the parts for this module (listed later)
> - You know a bit about computer programming (You have had at least had an introductory course.)
> - You know a resistor, from a capacitor, from a transistor, from an LED

© 2012 Robots-and-Androids.com

Arduino Module 1: Let There be (LED) Light!

Itinerary

1. A VERY Brief History of C
2. A VERY Brief History of the Arduino Microcontroller
3. Downloading and installing the Arduino IDE
4. Introducing the Arduino IDE
5. Lab 1: Hello World! with Blinky Lights
6. Lab 2: Run it Once – More Blinky Lights
7. Lab 3: Looping It – Even More Blinky Lights
8. Lab 4: Counting, Looping & Variables

A Request of Our Students:

- Please NOTIFY us by leaving a message on our web site (http://www.robots-and-androids.com) if you find mistakes of ANY type, if you are unclear of anything, and if you have suggestions, they are always welcome.

© 2012 Robots-and-Androids.com

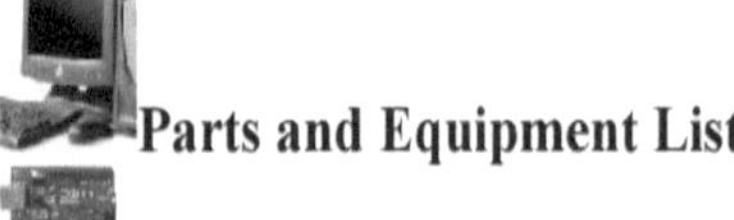

Parts and Equipment List

- 1 Windows PC running Microsoft Windows xP or better
- 1 Arduino Uno or Duemilinova Arduino microcontroller board
- 1 LED of any color

© 2012 Robots-and-Androids.com

What is a Microcontroller?

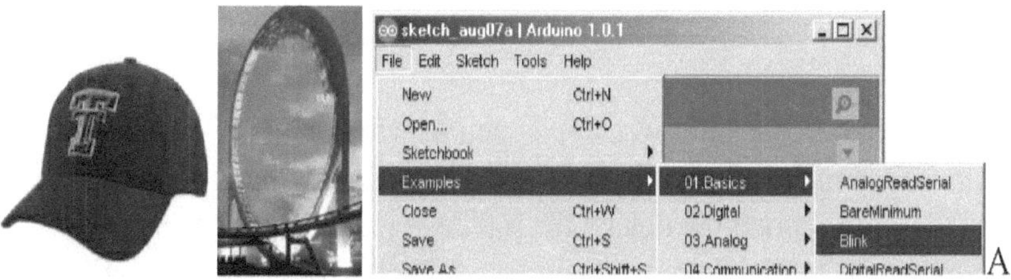

A microcontroller is a tiny computer usually on a single computer chip, usually without a screen. It often requires a full-size computer to program it. In addition to robots, microcontrollers are used in cars, TVs, remote controls, MP3 players, cell phones, and so most other electronic devices these days.

Gary Boone and Michael Cochran, engineers at Texas Instruments succeeded in creating the first microcontroller, the TMS 1000, in 1971. The result of their work was sold commercially in 1974.

A VERY Brief History of the Arduino Microcontroller

The Arduino microcontroller was developed in Ivrea, Italy in 2005. The intention was to make a very cheap (about $35 at Radio Shack at the time this was written.) programmable device for running student electronics projects. The inventors wanted a plug-and-play device. They wanted something someone could take out of a box, plug into a computer, and use immediately. Up to that time, microcontrollers were not so simple, nor so cheap. The favorite microcontroller used in many colleges before the Arduino was the Basic Stamp Microcontroller which goes for about $50 and requires a number of additional parts.

The Arduino was named after the inventors' favorite pub, the "Bar de Arduino" which in turn was named after Arduin of Ivrea, an ancient king of Italy. (There is more on the Wikipedia, but we wouldn't want to bore you.)

The Arduino, because of its price and ease of use then became very popular. There is even an Arduino called the Lilypad meant to be attached to clothing. This especially attracted a lot of ladies to Arduino.

A VERY Brief History of the C Language

The C programming language was developed between 1969 and 1973 by Dennis Ritchie at the Bell Telephone Laboratories for use with Unix.

It was named "C" because its features were derived from an earlier language called "B", a stripped down version of BCPL, Basic Combined Programming Language. Which had nothing to do with BASIC. (There is more on the Wikipedia, but again, we didn't want to bore you.)

Arduino Uno or Duemilanove?

There are two main types of Arduino boards floating around in the "Arduinosphere," the Duemilanove and the Uno. "Duemilanove" means "2009" in Italian and is named after the year of its release. "Uno" means "one" in Italian and is named to mark the release of Arduino 1.0. software.

Which one should you use? The Duemilanove, or "Duey" as we like to call them, was one of the first Arduino releases and is a solid microcontroller. Unfortunately it is no longer being supported. The Uno is the official latest release.

By the way, there are other Arduino models out there, but these two seem to be 99% of what people buy.

The Arduino IDE

The Arduino IDE (**I**ntegrated **D**evelopment **E**nvironment) is where you write
your programs, compile your programs (Compile: turn programs into
programs which the Arduino can understand), Upload your programs (move
them from your PC to your Arduino), and monitor your programs.
Essentially, it is an editor, a compiler, an uploader, and it holds your of
Arduino programs and sample programs among other things.

Sketches? Sketches? But Where are the programs?

As you may have already heard, Arduino programs are called "sketches" and
they are kept in a folder called a "sketchbook" Quaint? Yes?

© 2012 Robots-and-Androids.com

Downloading and Installing the Arduino IDE

To write Arduino "sketches" (programs) you will need to download and install the Arduino IDE, the Arduino Integrated Development Environment).

You will need to go to the Arduino web site (http://arduino.cc/en/Main/Software) and download the IDE to your desktop. The version that was used when we wrote this eBook was 1.0.1. This is bound to change. Also, we wrote this with the PC in mind, but it should work just as well with a Mac or Linux computer.

After you get to the Arduino web site, in the "download" section, select your computer and you will be asked if you want to download the IDE. Select your desktop as the place to download it and select "save". It's a pretty big "zip" file and may take quite a few minutes to download.

After the IDE has competed downloading, you will need to install it.
1. Double click on the zip file that you just downloaded.
2. Drag out the Arduino-1.0.1 (current version) folder from the zip file and put it on your C:/ (or C:/programs/) folder. Windows 7 and some Vista versions won't let you put anything at the C:/ level.
3. You might have to wait a few minutes for the file to unzip. Again it is HUGE.
4. Drag the file called *arduino.exe* to your Windows Start menu, or toolbar, or desktop.
5. And you're Arduino IDE is ready!

An Important Note

Please be aware that the current Arduino revision which at the writing of this eBook was revision 3, ("rev3" on the boards) requires that the USB drivers be MANUALLY INSTALLED. This is a royal pain in the Arduino. See our web site if you need help on installing the USB drivers. (http://www.robots-and-androids.com)

Please note, with the release of the Arduino Uno came a problem with it finding the right USB port. The UNO has built-in software to find the right one itself. It is often wrong. It sometimes will take a second guess, asking you, the user if it is alright to change ports, and then (hopefully) selecting the right port. Unless the right port is connected, your programs ("sketches" in ArduinoSpeak) will never be uploaded to your Arduino.

It is best to remove all other USB hardware (including memory sticks) from your PC and run the Arduino by itself. You will be more successful this way. We found out that if other USB products are running concurrently with the Arduino, they will stop the Arduino from finding the correct USB port. USB hubs were especially problematic.

Get Ready to Rock!

1. Turn off your PC if it is already on.
2. Plug the USB cord that came with your Arduino into the Arduino board.
3. Plug your Arduino USB cord into a USB port on your computer, making sure there are **no other USB items plugged in**.
4. (Although there is a place to plug in a 6 volt power supply, for this first module you don't need to plug it into any power since it will take its power from your computer. When an Arduino is run without a computer, the Arduino board requires between 7 and 12 volts. For now though, don't use an external power supply.)
5. Start up your PC.

If you always use the same USB port, you will probably only need to start your Arduino this way the first time. After that you should be able to just plug it in (into the same port as you used before) and go. But beware, Arduinos have been known to change their port number without warning!

Introducing the Arduino IDE

Now that you have installed the Arduino IDE, it is time to start it up. Go to your desktop (or into the Arduino folder in your C:\\ drive) and double-click on Arduino101.exe.

After the Arduino splash screen goes away, you should see an Arduino IDE screen something like the one in the photo.

Let's spend a minute introducing you to some of its features.

The Arduino Menus

It has five main menus: File, Edit, Sketch, Tools, and Help.

File: This is where you can create a *New* sketch, *Open* an old sketch, look at YOUR *Sketchbook* (the programs that you write and save), and look at the *Examples* that come with the IDE.

Take a quick look at all the examples that you have in the IDE. (File / Examples) As you can see, there are more than enough to get you started. And if you bother to search online, you will find thousands of more examples. You will see that, Arduino has become QUITE popular as a microcontroller and there are THOUSANDS upon THOUSANDS of programs out there just waiting for you to try.

Edit: This is like most word processors with cut, copy, paste, and more.

Sketch: You will use this mostly to select Verify/Compile.

Tools: This is where you will select your Arduino board and serial port. More about that later.

Help: You will of course use this when you are stuck and need a bit of help.

A real nice feature within the IDE is the *Find in Reference* menu item. You can highlight the Arduino command that you need help with and select *Find in Reference* from the menu, OR you can just highlight the Arduino command that you need help with and do a <CTRL+SHIFT+F> and it will open the help files to the correct reference.

© 2012 Robots-and-Androids.com

The Arduino Icons

Now let's check out the six icons on the IDE.

-  Verify (checkmark): used for making sure you did not make any mistakes.
- Upload (right arrow): used for sending your sketch from your PC to your Arduino board. (Upload will also verify before it sends out your sketch.)
- New (sheet of paper): used for creating a new sketch.
- Open (up arrow): used for opening an existing sketch.
- Save (down arrow): used for saving your changes to a sketch.
- Serial Monitor (magnifying glass): used for monitoring your Arduino programs as they run.

© 2012 Robots-and-Androids.com

Lab 1: Hello World! with Blinky Lights

In any computer language class, your first program is usually called "Hello World!" This program is to prove to yourself and the rest of the world that yes! you can upload and run a simple program. Here at Robots and Androids we just LOVE blinking lights on robots! We also love reciting pi in binary to a thousand digits for fun. We are just of that mindset.

Hardware: There is no special hardware for this first sketch. If your Arduino is plugged in and its little onboard LED is on, you are set.

Software

1. Open the sample sketch called *Blink*. You can find it in your IDE here: File / Examples / Basics / Blink (Note: in the computer world, your sketch (your program) is often known as "source code" or just "code" for short. We will use these terms interchangeably)
2. Click the "Verify" button (checkmark icon) to make your sketch into a format that the microcontroller understands. Your IDE should not report any errors, and it should report that your sketch was compiled successfully. This is also called "compiling your code." You now see that "Verify" does more than verify, it compiles!
3. Upload the Blink sketch to your Arduino board by clicking the "Upload" button (right arrow icon). Your TX and RX lights on your Arduino board should start flashing rapidly. When they stop, the IDE reports that your upload was successful.
4. One of your onboard LEDs (there are 4 onboard LEDs on most Arduinos) will start flashing once per second.
5. Congratulations!!! You have PROGRAMMED A MICROCONTROLLER!!! You are now well on your way to becoming a famous roboticist! (Don't forget the "Three Laws of Robotics". As before, more can be found on Wiki.)

Now for a bit of excitement (well perhaps not that much excitement) take a standard LED (blue is a good color) and plug the shorter leg into GND and the longer leg into digital pin 13. If you have the polarity of the LED correct you will see that it flashes in sync with the LED that is mounted on the Arduino board. This is fun! (Right?)

Dissection of the Sample Program Called "Blink"

You might be asking yourself "So what exactly do all those strange words in the sketch mean? How does it work? Why do I need the word 'void'? What will I have for supper?" and other such things. Let's start at the top and explain it all line-by-line. Below is the sketch (or code) dissected piece-by-piece.

```
/*
  Blink
  Turns on an LED on for one second, then off for one second,
repeatedly.

  This example code is in the public domain.
*/
```

As you can see, the print is in gray. Whenever you start off with a slash followed by an asterisk (/*), you are telling the IDE to IGNORE all of the following text.

On the last line you see an asterisk followed by a slash(*/). This tells the IDE to STOP ignoring the text and pay attention to the rest of what you have written. If you were to forget and leave the asterisk slash out, the IDE might try to ignore your entire sketch, but probably will give you an error.

```
// Pin 13 has an LED connected on most Arduino boards.
// give it a name:
```

Now you see that the text starting with "Pin 13" is gray". The double slashes (//) tell the IDE to ignore only the line that the slashes are on, and only from that point forward. You can use either "//" to ignore single lines (or parts of a line) or "/* */" to ignore multiple lines of text. It is up to you. These are used for both leaving comments in the code for documentation purposes as well as "hiding" code from the IDE that you don't wish to run.

```
int led = 13;
```

"int" means we are telling the IDE that we wish to use a variable called "led" as an integer and that it will have a value of "13". There is a semicolon at the end of the line. The C language requires semicolons at the end of most lines of code. Notice that "int" is colored orange? The orange coloring indicates that it is a reserved word, that is, a word that has a special meaning. You would not be able to use "int" as a variable name.

```
// the setup routine runs once when you press reset:
```

More comments. These comments tell you what "setup" does, and that is quite right. All code between the two curly brackets after the command "setup" (see the code below) will be run just once.

```
void setup() {
        // Your code goes HERE
    }
```

Here we begin the code that the microcontroller will only run once.

"Setup" requires: 1. the word "void" before it, 2. an open and closed parenthesis after it, and 3. all your setup code between the left and the right curly brackets. Why does it need the word "void"? You can look that up yourself.

```
pinMode(led, OUTPUT);    // initialize the digital pin as an output.
```

Here, as the comment says, you art telling the Arduino board that pin 13 (that is the LED's pin which is connected to pin 13 on the board which you defined earlier as an integer) will be used as an output. In other words, you will later put 5 volts on pin 13 which will turn on the LED. If you wanted to **read** a voltage on it, you would set it as and **input**.

```
    }
```

A right curly bracket signals the end of your "run once" code.

```
// the loop routine runs over and over again forever:
void loop() {
        // Your code goes HERE
    }
```

As the comment says, "loop" runs forever and ever. In the robot world this is a good thing. As with "setup", "loop" also requires "void", an open and closed parenthesis, and curly brackets to start (and end) the loop.

```
digitalWrite(led, HIGH);   // turn the LED on (HIGH is the voltage
level)
```

As the comment says (and yes, you can put a comment in on the same line as code that will run) this line tells the Arduino to put a "high" voltage (a voltage of about 5 volts) on line 13 (called "led"). "digitalWrite" is used to command the microcontrollers pins to do various things. In this case it is turning on the voltage, which will then light the LED.

```
delay(1000);               // wait for a second
```

As the comment says, we are delaying the sketch for a time, 1000 milliseconds in fact. 1000 milliseconds equals one second. If we did not delay the code, our sketch would run so fast that you would never see the LED come on.

```
digitalWrite(led, LOW);    // turn the LED off by making the voltage
LOW
```

As the comment says this line tells the Arduino to put a "low" voltage (a voltage of about 0 volts) on pin 13. "digitalWrite" is again used to command the microcontroller to turn off the voltage, which will then shut off the LED.

```
delay(1000);               // wait for a second
```

Same as the other delay. If we did not have this, the LED would look like it was never shut off.

```
}
```

As the other right curly bracket signaled the end of your "run once" code, this one signals the end of the loop. All the lines of code that you wish to run again and again should be placed before this final curly bracket.

And that ends your sketch. In the next lab we will make some changes and customize the blink.

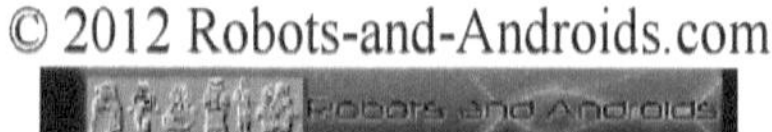

Lab 2: Run it Once – More Blinky Lights and a Counting loop

Now that you have compiled, uploaded and ran your first Arduino sketch, it's time to do a bit of changing things around.

First thing you need to do is copy the following lines of code and paste them BEFORE the right curly bracket in the "setup" section so that it looks something like the code below:

```
// the "setup" routine runs once when you press reset:
void setup()
     {
     pinMode(led, OUTPUT);      // initialize the digital pin as an
output.

     digitalWrite(led, HIGH);   // turn the LED on (HIGH is the
voltage level)
     delay(1000);               // wait for a second
     digitalWrite(led, LOW);    // turn the LED off by making the
voltage LOW
     delay(1000);               // wait for a second
     }
```

Warning: make sure that all the copied code goes BEFORE the right curly bracket or your sketch will NOT work. Also, make sure you do not accidentally delete the curly bracket or any of the semi-colons or, again, your sketch will NOT work.

Next you will need to put a comment symbol, the double slash "//" before each line of the code inside of the loop section. You will notice that the code turns gray after you put the slashes on the line. This means your IDE will ignore the code. The copied code in the setup section will run, but not the code in the loop section. It should look something like this:

```
// the "loop" routine runs over and over again forever:
void loop()
     {
```

```
            //digitalWrite(led, HIGH);    // turn the LED on (HIGH is the
    voltage level)
            //delay(1000);                // wait for a second
            //digitalWrite(led, LOW);    // turn the LED off (LOW is the
    voltage level)
            //delay(1000);                // wait for a second
        }
```

Warning: make sure you do not put just a single slash "/" in front of your code in the loop section. You need two of them or the sketch will NOT run. Also be sure that you do not put slashes in front of the void loop () { line, nor in front of the final right curly bracket.

1. Again, click the "Verify" button (checkmark icon) to make your sketch into a format that the microcontroller understands. Your IDE should not report any errors, and it should report that your sketch was compiled successfully. This is also called "compiling your code." You now see that "Verify" does more than verify. See picture below.
2. And again, upload it to your Arduino board by clicking the "Upload" button (right arrow icon). Your TX and RX lights on your Arduino board should start flashing rapidly. When they stop, the IDE reports that your upload was successful.
3. Your onboard LEDs (there are 4 onboard LEDs on most Arduinos) will flashing once.

"Once?" you may ask. Yes, only once. Since you flashing code is now in the setup section, it only will run once.

Click the small button on your Arduino board. Ah! Your code runs again. That button is the "reset" button. Pushing it starts the code over again.

Click File / Save, and save your sketch as "*sketch1*" or something like that. You cannot overwrite the Blink sample code. The IDE forces you to save it under a different name.

Now uncomment (remove the double slashes) from the code in the loop section.

Change the delays to something more or less than is there. Try some numbers between 100 or 4000 in your delay commands. Then upload that sketch and see your result. Is it what you expected? Make a few more changes and run as you wish.

Lab 3: Flashing in Sequence – Even More Blinky Lights

The next thing you are to do is to move the LED that you added earlier to a different pin. Keeping the shorter lead attached to GND, plug the longer lead into digital pin 12 (yup, right there next to 13.). Next add the following to your setup code:

```
pinMode(12, OUTPUT);
```

Your sketch should look something like this:

```
void setup()
    {
    pinMode(led, OUTPUT);    // This is for the LED built into
the Arduino.
    pinMode(12, OUTPUT);    // The LED that you moved,
initialize digital

                            //pin 12 as an output so it lights up
                    separately

    digitalWrite(led, HIGH);   // turn the LED on (HIGH is the
voltage level)
    delay(1000);               // wait for a second
    digitalWrite(led, LOW);   // turn the LED off by making the
voltage LOW
    delay(1000);               // wait for a second
    }
```

But that is not enough. You will need to turn the LED on pin 12 on and then off just like the built-in LED on pin 13. So, add digitalWrite and delay statements to do this so that your code kind of looks like this:

```
void loop()
    {
    digitalWrite(led, HIGH);   // turn the LED on (HIGH is the
```

```
        voltage level)
        delay(1000);              // wait for a second
        digitalWrite(led, LOW);   // turn the LED off by making the
voltage LOW
        delay(1000);              // wait for a second

        digitalWrite(12, HIGH);   // turn the LED on (HIGH is the
voltage level)
        delay(1000);              // wait for a second
        digitalWrite(12, LOW);    // turn the LED off by making the
voltage LOW
        delay(1000);              // wait for a second
        }
```

Now upload and run your sketch. What you should see is the built-in LED light for a second, then the other LED light for a second followed by the built-in again. This of course, because it is in the "loop" section, will run endlessly—at least until you turn it off or upload a different sketch.

Notice that in one set of digitalWrite statements, we used the variable "led" while in the other we just used the number 12? As a programmer, you can do either. The preferred way is to use variables. We could of used a second variable called "led2", but we got lazy and figured we could just type in the number "12". You can replace the "12" with a new variable if you wish. We won't stop you. Just don't forget to initialize it in the setup section the same way that "led" is initialized. (This should work: int led2 = 12;)

Lab 4: Counting, Looping & Variables – Counting Blinky Lights!

The Variables of the Arduino C language.

The Arduino C language uses many types of variables. It represents these using various abbreviations.

The C language uses:
"int" for integers such as 1, 2 , or 9999.
"float" for numbers with a decimal point such as 3.14
"char" for letters and words such as "a" or "apple"

This is just a short list. There are many others.

In Arduino C, (and most other C languages) you must tell the IDE what type of variable it is before you actually use it. This is called "declaring" your variable.

For those not too familiar with variables, we like to think of variables like cups that you can store things in. For example if you have an integer cup called "myInteger" you can store an integer in it such as "99". You do this like this:

    int myInteger = 42;

In this way you set the value at the same time that you declare it. Kind of convienient that way.

You can also change it later on like this:

    MyInteger = -33;

Or

    MyInteger = YourInteger;

Assuming that you have a second variable called "YourInteger" that has also been declared as an integer.

In this section we take this idea a bit further. We will use variables within a loop for counting how many times we've looped within that loop.

This counting loop is called the "For" loop. It goes something like this:

```
for (int x = 0; x < 5; x++)
    {
        // your code goes here!
    }
```

One more time, open the "Blink" sample code in a new window. Close all your other open Arduino IDE windows so that only this new one is open.

Enter your "For" loop code in the top setup section. Copy the lines of code that turn the built-in LED code on and off between the curly brackets of the "for" loop. Your code should look something like this:

```
// the setup routine runs once when you press reset:
void setup()
    {
    pinMode(led, OUTPUT);    // initialize the digital pin as an
output.

        for (int x = 0; x < 5; x++)
            {
                digitalWrite(led, HIGH);   // turn on the LED
                delay(1000);               // wait for a second
                digitalWrite(led, LOW);   // turn off the LED
                delay(1000);               // wait for a second
            }
    }
```

Notice that there are TWO sets of curly brackets? One is for the setup section and the other is for your for loop.

Again comment out (put slashes in front of) the digitalWrite and the delay code in the loop section of the sketch so that it looks something like this:

```
void loop()
    {
    //digitalWrite(led, HIGH);   // turn on the LED
```

```
//delay(1000);            // wait for a second
//digitalWrite(led, LOW);   // turn off the LED
//delay(1000);            // wait for a second
}
```

Note that you cannot just delete the loop code out of your sketch. If you do, your IDE will display an error notice and your sketch will NOT work. All sketches require a setup section and a loop section. To leave out either will stop your sketch from running.

Now make changes to the count, and the delay. Go ahead and add a second LED as you did before. Go ahead and add a third LED if you dare. Go ahead and make lots of changes and see how each one affects your sketch. With a little luck your Arduino will start looking like its own little light show and that's a good thing. You learn by doing, so do a lot and learn a lot.

—Thomas Messerschmidt,

© 2012 Robots-and-Androids.com

Recap:

Well, now you can certainly call yourself a C programmer, and even an Arduino programmer. I hope you have learned something here in this online workshop. Here is a summary of what we covered:

1. The history of microcontrollers, the C programming language, and the Arduino
2. Working with the Arduino IDE
3. Compiling and uploading sketches
4. Variables
5. Loops
6. Counting loops
7. Modifying sample code to have it do what YOU want it to
8. Adding hardware (an LED) to your Arduino board.
9. When you start building your robot, you now know how to get all sorts of cool LEDs flashing on it.

© 2012 Robots-and-Androids.com

We have also learned the following Arduino C code and Arduino reserve words:

Symbol / Reserved Word	Meaning
/*	Start a comment
*/	End a comment
//	Start a one-line comment or tells the IDE to ignore the text or code that follows it.
;	Used at the end of most C statements.
int	Used to declare that a variable will be an integer.
setup	Used to show code that will only run once.
void	A necessary part of "setup" and "loop".
{	Used to show where "setup" and "loop" code starts. Used also for other loops such as "for", as well as decisions.
}	Used to show where "setup" and "loop" code ends. Used also to end other loops such as "for", as well as decisions.
pinMode	Used to tell the microcontroller how a pin will be used.
OUTPUT	An indication that a microcontroller pin will have some voltage on it.
Loop	Used to show code that will run again and again (forever)
digitalWrite	Used to tell the microcontroller to put a voltage on a pin or shut it off.
delay	Used to stop the code for some length of time. (in microseconds.)
HIGH	A voltage of about 5 volts.
LOW	A voltage of about 0 volts.

© 2012 Robots-and-Androids.com

The THREE Secrets to Great Arduino Programming:

1. BACK UP YOUR WORK!!!
You never can tell when things will go wrong, go wrong, go wrong…

2. COMMENT YOUR WORK!
You'll want to know why you did what you did some weeks, months, years
from now.

3. BACK UP YOUR WORK!!!
You never can tell when things will go wrong
go wrong, go wrong…

(This list is subject to change.)

© 2012 Robots-and-Androids.com

Additional Modules to Come

- The Capacitive Touch Sensor
- Arduino Music
- The Capacitive Touch Musical Organ
- The Singing Arduino ("Daisy, Daisy…")
- The Talking Arduino
- Arduino Add-on Hardware: Shields
- Blue Tooth Shield
- LCD Shield
- I2C Communication
- Accelerometers
- Gyros
- Motor Control: The H-Bridge
- More Touch Sensors
- Building Battle Robots!
- Building Robot Grippers (end effectors)
- Learning Arduino C : The Basics
- Learning Arduino C : Intermediate
- Learning Arduino C : Advanced
- Learning Arduino C : Bits, Bytes, and Registers
- Learning Arduino C : Sensors
- Learning Arduino C : Motor Control

(This list is subject to change.)

© 2012 Robots-and-Androids.com